LIFE SCIENCES CONTRIBUTIONS

ROYAL ONTARIO MUSEUM

NUMBER 106

C. R. BARNES
D. J. SLACK

Conodont Ultrastructure: The Subfamily Acanthodontinae

Publication date: 20 October 1975

ISBN: 0-88854-179-1

Suggested Citation: Life Sci. Contr., R. Ont. Mus.

ROYAL ONTARIO MUSEUM PUBLICATIONS IN LIFE SCIENCES

The Royal Ontario Museum publishes three series in the Life Sciences:

LIFE SCIENCES CONTRIBUTIONS, a numbered series of original scientific publications, including monographic works.

LIFE SCIENCES OCCASIONAL PAPERS, a numbered series of original scientific publications, primarily short and usually of taxonomic significance.

LIFE SCIENCES MISCELLANEOUS PUBLICATIONS, an unnumbered series of publications of varied subject matter and format.

All manuscripts considered for publication are subject to the scrutiny and editorial policies of the Life Sciences Editorial Board, and to review by persons outside the Museum staff who are authorities in the particular field involved.

CHRISTOPHER R. BARNES is a Research Associate, Department of Invertebrate Palaeontology, Royal Ontario Museum, and an Associate Professor in the Department of Earth Sciences, University of Waterloo, Waterloo, Ontario.
DOUGLAS J. SLACK is a recent Honours graduate of the Department of Earth Sciences, University of Waterloo, Waterloo, Ontario.

PRICE: $1.50

100 Queen's Park, Toronto, Canada

PRINTED AND BOUND IN CANADA AT THE JOHN DEYELL COMPANY

Conodont Ultrastructure: The Subfamily Acanthodontinae

Abstract

The ultrastructure of representatives of the condont Acanthodontinae has been studied with the aid of the scanning electron microscope. Five form species of *Scolopodus* and one of *Ulrichodina* are the most common taxa of this subfamily in those faunas from which material was selected: St. George Formation, northern Newfoundland; Baumann Fiord Formation, Ellesmere Island; and Mystic Formation, southern Quebec. These species are typical of Lower and early Middle Ordovician faunas of the Midcontinent Province. Etched longitudinal and transverse sections were studied to determine internal structure for comparison with external micromorphology. The cone-in-cone lamellar structure is modified in certain taxa through secretion of localized lamellae that lengthen the element distally and/or proximally and construct prominent carinae and costae. In other taxa, fine external striae can be detected on internal lamellae representing early ontogenetic stages. Two species (*S. emarginatus*, *S. gracilis*) display radial lamellae supporting a close relationship with the Panderodontidae. Most structures of the elements parallel the postero-anterior plane and may be strengthening adaptations in that plane for these simple cones.

Key Words: conodont; ultrastructure; Acanthodontidae; *Scolopodus*; *Ulrichodina*; Ordovician.

Introduction

The study of the internal structures of conodonts has greatly increased with the advent of the scanning electron miscroscope (SEM). High magnifications and various sectioning techniques used by recent workers (e.g., Barnes, Sass, and Monroe, 1970, 1973; Barnes, Sass, and Poplawski, 1973; Lindström and Ziegler, 1971; Lindström et al., 1972; Müller and Nogami, 1971; Pierce and Langenheim, 1969, 1970; Pietzner et al., 1968) have provided new data to assist in the taxonomic classification of conodonts. For example, Lindström's (1970) proposed suprageneric classification is partially based on the micromorphology of the conodont elements.

It might be assumed, yet needs to be verified, that those taxa with similar surface micromorphology are closely related. It is not known to what extent surface features persist through all growth stages and thus may be preserved in the interior of the conodont elements. Even the basic cone-in-cone pattern of conodont growth appears to be more complex than earlier believed. For instance, a different form of lamellae, radial rather than concentric, has recently been detected in certain taxa, and there is now evidence that particular concentric lamellae did not extend over the entire element (Barnes, Sass, and Poplawski, 1973) in the manner commonly depicted (e.g., Müller and Nogami, 1971).

In a recent study of one of the most common Lower Palaeozoic simple cone genera, *Panderodus*, both concentric and radial lamellae were noted (Barnes, Sass, and Poplawski, 1973). The inner lateral surface of specimens of *Panderodus* is characterized by longitudinal striae, subparallel to the longitudinal furrow, that are the surface expression of the radial lamellae. It seemed possible, therefore, that other simple cones exhibiting marked surface striations might possess radial lamellae. A further hint of this possibility was provided by a vague radial pattern shown on the fractured surface of a specimen of *Scolopodus gracilis* (Barnes and Poplawski, 1973, pl. 3, fig. 7A). Thus, this species and others of *Scolopodus* were investigated to determine their internal structure. A few other taxa in Lower Ordovician faunas also exhibit surface striations. One of the most common of these is *Ulrichodina*, and the type species, *U. prima*, was also selected for examination. This paper thus summarizes the results of the ultrastructure studies of five species of *Scolopodus* (*S. cornutiformis*, *S. emarginatus*, *S. gracilis*, *S. multicostatus*, *S. quadraplicatus*) and one of *Ulrichodina* (*U. prima*).

Taxonomic Status of Scolopodus and Ulrichodina

A current revolution in conodont taxonomy involves supressing, where possible, form taxa in favour of multielement taxa. Factors involved in the erection of such taxa include surface micromorphology and the presence, abundance, and distribution of white matter that is usually located in the distal portions of the cusp and denticles (Lindström, 1964, pp. 17–22).

In his recent revision of certain Lower Ordovician conodonts into multielement taxa, Lindström (1971, pp. 40–41) emended the genus *Scolopodus* to include "hyaline, drepanodiform elements with rounded cross-section and symmetrical as well as asymmetrical elements. The sides of the elements may be finely costate. The base is not expanded greatly". In his work with Baltoscandian conodonts, Lindström (1955, 1971) encountered only *Scolopodus rex* Lindström, which comprises a minor proportion of those and other faunas of the North Atlantic Province. However, other species of *Scolopodus* are dominant components of contemporaneous faunas of the Midcontinent Province (e.g., Barnes, Rexroad, and Miller, 1973; Bergström, 1973). This province, which is represented in North America, Australia, and Siberia, is interpreted as being developed in epeiric seas with high temperatures and salinities located in tropical latitudes (Barnes and Fåhraeus, 1975). The North Atlantic Province is regarded as a fauna occupying normal marine environments.

While *S. rex* of the Baltic area is hyaline, i.e., virtually devoid of white matter, the species examined in this study possess variable amounts of white matter. In most, this was confined to the central growth axis and adjacent areas. However, in *S. cornutiformis* white matter is present generally in the distal three-quarters of the cusp. Thus, the generic definition proposed by Lindström (1971) will probably require emendation. Perhaps the erection of a new genus for the scolopodids discussed herein is justified, but this must await a detailed study of large collections.

It is possible that the scolopodids of the European and Midcontinent provinces were closely related during the early Arenigian but became separated with increasing provincialism during that epoch. Alternatively, the scolopodids of the two provinces may be taxonomically different at a suprageneric level. Lindström (1970, pp. 231, 233) advocated this view in bringing *Scolopodus* Pander (including *Scolopodus rex*; Lindström, 1971) to the Oistodontidae of the Chirognathacea, while those scolopodids typical of Midcontinent faunas were assigned to the Acanthodontidae of the Panderodontacea (Table 1). Although not listed by Lindström (1970), it would appear that *Ulrichodina* also belongs in the Acanthodontidae (Acanthodontinae) on the basis of its morphology, micromorphology, and similar geographic and stratigraphic range, and is thus included in this paper. In previous studies, the only representatives of this subfamily examined by Lindström and Ziegler (1971) in their study of ultrastructure in the Panderodontacea were single specimens of the form species *Acontiodus iowensis* Furnish and cf. *A. staufferi* Furnish. The detailed ultrastructure of the closely related Panderodontidae has been described by Barnes, Sass, and Poplawski (1973); that of the Chirognathidae was discussed by Barnes, Sass, and Monroe (1973).

Table 1. Part of Lindström's (1970) suprageneric classification of conodonts. Scolopodids have been placed in both the Oistodontidae and Acanthodontinae.

Superfamily Chirognathacea (Branson and Mehl, 1944)
- Family Oistodontidae Lindström, 1970
- Family Chirognathidae Branson and Mehl, 1944
- Family Rhipidognathidae Lindström, 1970

Superfamily Panderodontacea Lindström, 1970
- Family Acanthodontidae Lindström, 1970
  - Subfamily Acanthodontinae Lindström, 1970
  - Subfamily Protopanderodontinae Lindström, 1970
- Family Panderodontidae Lindström, 1970

Materials and Methods

The specimens were obtained primarily from the St. George Formation (Lower Ordovician) of northern Newfoundland, with supplemental material from the Mystic Formation (Middle Ordovician) of southern Quebec and Member B of the Baumann Fiord Formation (Lower Ordovician) of central Ellesmere Island, Arctic Canada. They were part of faunas that have been discussed by Barnes and Tuke (1970), Barnes and Poplawski (1973), and Barnes (1974) respectively.

The studied specimens represent six form species within two form genera as outlined in Table 2. All the specimens used for internal ultrastructure studies were first embedded in Bioplastic (Ward's, Rochester, N.Y.) in the desired orientation and then ground down with a carborundum grinding wheel. They were buffed to a high polish with the use of a felt disc and tin oxide polishing compound, and finally etched with 2N HC1 for approximately 30 seconds before being coated and viewed in the Cambridge Stereoscan Mk. IIa scanning electron microscope at the University of Waterloo.

Both longitudinal and transverse sections were prepared for each species. Other sections present oblique views or areas around the basal cavity. Several good specimens were sectioned serially by repeated grinding. Seventy specimens (Table 2) were used, the majority for examination of surface micromorphology but with many selected for internal studies, and from the latter group 27 sections were prepared.

All specimens are deposited in the Department of Invertebrate Palaeontology, Royal Ontario Museum, Toronto, Ontario.

Table 2. Taxa used in study.

Localities and Species	No. of Specimens	No. of Sections
St. George Formation, Newfoundland, (L. Ord.)		
Scolopodus cornutiformis Branson and Mehl (1933)	9	4
S. emarginatus Barnes and Tuke (1970)	12	4
S. gracilis Ethington and Clark (1964)	4	2
S. multicostatus Barnes and Tuke (1970)	7	2
S. quadraplicatus Branson and Mehl (1933)	28	8
Ulrichodina prima Furnish (1938)	5	2
Mystic Formation, Southern Quebec, (Mid. Ord.)		
U. prima Furnish (1938)	1	1
Member B, Baumann Fiord Formation, central Ellesmere Island, (L. Ord.)		
S. gracilis Ethington and Clark (1964)	2	2
U. prima Furnish (1938)	2	2
Total	70	27

Ultrastructure of Acanthodontinae

The cusp of most conodont elements is constructed of two types of material: hyaline (lamellar) material and white matter (Lindström, 1964). Both the hyaline material and the white matter are similar in chemical composition (see Pietzner *et al.*, 1968). Hass and Lindberg (1946, p. 503) concluded that the crystallographic axes and the optic axes of the individual crystallites are coincident, and that the "crystal units in each lamella of a conodont are oriented in conformity with the direction in which the conodont grew".

Hyaline and White Matter

The hyaline matter of the conodonts considered herein is composed of innumerable microscopic crystallites arranged into lamellae (Figs. 5D, E; 6C, D, E). When viewed in longitudinal section, the lamellae appear to be composed of crystallites arranged end-to-end with the long axes (c-axes) oriented almost parallel to the long axis of the conodont (Fig. 5D, E). In cross-section, the crystallites appear to be oriented side by side, with their c-axes perpendicular to the plane of the section, to form concentric lamellae (Fig. 6C, D, E). Each lamella is separated from those adjacent to it by irregular interlamellar spaces but is fused to them irregularly along its length and width (Figs. 4B; 5E; 6D, E). The interlamellar spaces of the studied specimens are not well defined and are usually only inferred from a linear arrangement of voids paralleling the lamella. The etching process may have enhanced their definition. No basic difference, however, was noted in hyaline matter as described by previous workers, nor in the size of crystallites and lamellae as reported by Lindström and Ziegler (1971) for specimens of the Acontodontinae.

White matter is an opaque white material of virtually identical chemical composition to the hyaline material, but it differs in several respects. It does not display the regular crystallite arrangement of the hyaline material, but appears virtually structureless. It is, however, permeated by numerous irregular voids (Figs. 3D; 5D, E). These holes are small (less than 0.5μm) and apparently are randomly distributed throughout the white matter. The larger, somewhat linear voids do, however, show a very crude alignment, roughly parallel to the interlamellar spaces of the hyaline matter, and may indeed be the remnants of these interlamellar spaces.

White matter is generally accepted as being a secondary product of reorganized hyaline material. If so, one would expect to find the two substances in the process of recrystallization. This process is observable in a transition zone along the boundary of the two. The transitional material has been termed incipient white matter by Barnes, Sass, and Monroe (1973). Fig. 5D, E shows this transition in a longitudinal section of *S. quadraplicatus*, revealing that the transition begins with the appearance of small holes in the crystallites. The interlamellar spaces close and form irregular lines of voids rather than consistent interlamellar spaces. The crystallites lose their hexagonal habit and form agglomerates. This process continues until all apparent crystallite structure is lost and the true white matter is formed. Fig. 1C, D shows the closure of the interlamellar spaces and the formation of irregularly spaced holes within the crystallites and lamellae prior to complete recrystallization.

Surface Micromorphology

In their study of the ultrastructure of the Panderodontacea, Lindström and Ziegler (1971) outlined six types of surface ornamentation on their specimens, namely: smooth surfaces, fine striae (<0.7 µm), coarse striae and ridges (>0.7 µm), basal wrinkles, longitudinal furrow, and denticles. The last feature (denticles) was not observed in this study, since only simple cones were examined. While basal wrinkles are absent, we found all of the other features described by Lindström and Ziegler (1971).

Smooth surfaces appear in localized areas on some of the specimens studied herein. *U. prima* displays this feature on the crest of the larger costae and along the anterior edge. *S. cornutiformis* has smooth surfaces along and close to the anterior edge and displays only a few fine striae posteriorly.

Abundant fine striae are found on *S. multicostatus, S. emarginatus, S. gracilis,* and *U. prima* (Figs. 2A, B; 3A; 6A, B), and to a lesser extent on *S. cornutiformis* and *S. quadraplicatus* (Figs. 1A; 5A). The overall pattern of striae is commonly continuous over the length of the conodont, as seen in Figs. 2A, B and 6A, B. However, individual striae seldom extend the entire length of the cusp, tending to merge and pinch out at irregular intervals along it.

Fine striae seem to be the surficial representation of the irregular edges of the outermost lamellae. The crystallites forming these lamellae are arranged in roughly parallel rows, thereby forming the fine striae. The fine striae pinch out or merge at some point along their length, because fewer crystallites are required to complete an individual lamella toward the tip of the cusp.

The coarse striae and ridges are represented here on specimens of *S. cornutiformis, S. quadraplicatus, S. gracilis,* and *U. prima* (Figs. 1A, B; 3A; 5A; 6A, B). They are formed by the orientation of lamellae within the cusp and vary from the narrow, sharp ridges and grooves of *S. cornutiformis* to the broad, shallow ridges of *S. gracilis*. The methods of building these ridge-groove systems will be discussed below.

Only one of the studied species displays a deep longitudinal furrow, *S. emarginatus* (Fig. 2E, arrow). The longitudinal furrow is a narrow and deep posterolateral groove. It extends inwards almost to the growth axis and runs along the posterior edge to near the tip of the cusp, where it has not yet developed. Lindström and Ziegler (1971) suggested that the furrow may represent a site of muscle attachment.

Internal Structures

Several noteworthy structures are displayed in the various specimens of this study (Table 3). Some of these structures, such as the growth axis and, to some extent, the distribution of white matter, are common to all. Others, such as keels, the longitudinal furrow, radial lamellae, and special concentrations of white matter, are restricted to particular species.

The ubiquitous structure is the growth axis. This is a cylindrical tube running

Table 3. External and internal features of the taxa studied.

Species	Smooth surfaces	Fine striae	Coarse striae & ridges	Longitudinal furrow	Radial lamellae	Percentage of white matter
Scolopodus cornutiformis	minor	few	present	absent	absent	ca. 75
S. emarginatus	none	abundant	absent	present	present	<10
S. gracilis	none	abundant	present	absent (?)	present	<10
S. multicostatus	none	abundant	absent	absent	absent	<10
S. quadraplicatus	none	few	present	absent	absent	<10
Ulrichodina prima	minor	abundant	present	absent	absent	<10

from the tip of the basal cavity (Fig. 1B) through the entire length of the cusp (Fig. 5C, arrow). It is usually located slightly anteriorly and can be located in most cross-sections as a hole (Figs. 1C; 3C), or as a localized cylinder of white matter in an otherwise hyaline cusp (Figs. 1E; 5B; 6C). This specific localization of the white matter in and closely adjacent to the growth axis is probably an adaptation on the part of the conodont animal to strengthen this vital area (Barnes, Sass, and Monroe, 1973).

Radial lamellae were first recognized and described in detail by Barnes, Sass, and Poplawski (1973) for the Panderodontidae. Certain comparable structures are evident in *S. emarginatus* and *S. gracilis* (Figs. 2E, F; 3B-D). In Fig. 2E, F the radial lamellae can be seen to be the internal expression of a series of coarse striae on the anterior surface of the cusp. The lamellae appear to run almost parallel to each other for the length of the section and then converge on a rather deep and narrow groove, i.e., the posterior furrow that extends inwards from the posterior margin. The lamellae become rather indistinct at this point, however, so that it is impossible to test the statement of Barnes, Sass, and Poplawski (1973) that the radial lamellae are related to the longitudinal furrow. Fig. 3B-D shows sections through *S. gracilis* which also display a type of radial lamellae. The features here, however, terminate in the posterior groove, but they also appear to be the internal expression of coarse surficial striae. The interesting feature is the association of the radial lamellae with a band of white matter running from the posterior groove to the anterior edge. Since the sections in Fig. 3B-D are all serial sections of the same specimen (with B being more apical) it is evident that the band is complete only in the apical portion of the cusp. The purpose of this feature is unclear, but it may be a strengthening feature.

U. prima (Fig. 6C, E) exhibits a posterior keel structure. It can be seen from the enlargement (Fig. 6E) that this lineation results from the alignment of rather large crystallites from the growth canal to the posterior margin.

Modes of Growth of Conodont Elements

Conodont elements consist of a basal cavity around which a series of cone-in-cone lamellae are added concentrically (i.e., externally). Each lamella of the specimens considered herein consists of a series of crystallites placed end-to-end and roughly parallel to the direction of growth of the cusp (Fig. 5E). The lamellae are not quite parallel to the growth axis (Fig. 2C, arrow), but are inclined to it by about 5–10 degrees (the angle becoming sharper apically). In certain taxa, the conodont achieves a maximum width in this manner and then continues to grow lengthwise (without adding extra width) by adding these lamellar cones *only* to the tip and upper portions of the cusp. This can be seen in Fig. 5D where the lamellae intersect the growth axis and then proceed toward the outer margin and terminate there. Contrary to earlier studies (e.g., Lindström, 1964; Müller and Nogami, 1971) in which it was thought that the lamellae simply thinned at the outer margin but were continuous over the whole element, the lamellae shown here terminate at the outer margin. This implies that the secreting tissues surrounding the element added material only to the areas of maximum growth. This selective growth is also illustrated in Figs. 2C, D and 4A, B, where the basal cavity has been deepened by the addition of material to the basal margin. In the first example, *S. emarginatus*, the lamellae are added almost perpendicularly to the direction of maximum extension (growth). The initial lamellae are continuous over the whole length of the conodont and simply curve inwards over the basal cavity. Later lamellae are added selectively to the basal margin and terminate at both the inner and outer margins of the basal extension. In the second case, *S. gracilis*, the lamellae are directed inward at the margin. Note here, however, that the crystallites which comprise the lamellae are still oriented roughly parallel to the direction of growth. This is to be expected, since the apatite crystallites are hexagonal in form and would therefore crystallize faster in the c-direction than in the a-direction, thus facilitating the thickening of the lamellae while minimizing the energy required. A further mechanism noted, which allows the conodont to grow preferentially in one direction, is the increase in the width of the interlamellar spaces. This has the added benefit of minimizing the amount of phosphate required for growth, as does the selective addition of material to only these areas where growth is required (Lindström, 1964).

At least two distinctive methods of growth are indicated by our cross-sections. The first is represented by *S. quadraplicatus* (Fig. 5B) in which prominent costae are developed by adding lamellae only to portions of an originally ovoid cusp. The juvenile conodont element originally had an ovoid cross section with concentric lamellae surrounding the central growth axis (Fig. 5B). At some later stage in ontogeny, secretion of lamellae of constant width occurred only in localized areas, thus building up a series of ridges and creating intervening grooves. Fig. 5B illustrates that the extreme outside lamellae pinch out as they run into a groove.

Earlier workers considered that lamellae extended over the entire surface of the conodont. The only exception noted by Müller and Nogami (1971) was partial lamellae in areas of regeneration. Barnes, Sass, and Poplawski (1973) illustrated lamellae restricted to the basal area of *Panderodus*, noting that much of the growth of the element is restricted to the basal area. In some scolopodids, areas of growth can be demonstrated to be restricted to both the basal area and the distal part of the cusp.

S. cornutiformis and *U. prima* on the other hand display a different mode of growth for costae (Figs. 1E, F; 5C). Both of these species have retained their costae since their earliest ontogenetic stages. This is best shown in Fig. 6C, where each lamella completely encircles the cusp and exactly mimics the previous lamella in form. Even fine surficial striae are distinguishable on the outer surfaces of internal lamellae (Fig. 6E). The rate of growth outwards from the growth axis is determined by the width of each lamella, and not by the addition of extra lamellae as in *S. quadraplicatus*. Thus, with the prominent posterior keel of *U. prima*, the distance from the growth axis (Fig. 6C, arrow) to the point of this keel is about three times the distance from the growth axis to the anterior edge. The ratio of the width of a lamella about two-thirds the distance from the growth axis is thus approximately 3 : 1 in the posterior direction.

Taxonomic Note

In our opinion, the taxa studied herein, belonging to *Scolopodus* and *Ulrichodina,* are closely related and can be appropriately referred to the Acanthodontinae (Lindström, 1970). However, considerable variation exists in the ultrastructure reflecting the construction of particular features. Thus some short cones have a simple cone-in-cone structure, whereas longer cones have added localized lamellae only at distal and/or proximal locations. In other forms, lateral ridges are constructed by localized secretion of lamellae. Radial lamellae only occur in certain species (*S. gracilis, S. emarginatus*) and are associated with the longitudinal groove or furrow. They appear to be associated with areas of incipient white matter. If all these taxa are indeed closely related, they show considerable plasticity in their ultrastructure. The presence of the distinctive radial lamellae does favour their close relationship to the Panderodontidae (Table 1), in which these have also been detected (Barnes, Sass, and Poplawski, 1973).

One feature of this classification is the use of criteria such as white matter to distinguish between the scolopodids of the North Atlantic and Midcontinent provinces. These groups of scolopodids were placed in the Oistodontidae and Acanthodontidae respectively. White matter seems to be notoriously variable in certain Ordovician simple cones. Such variable distribution was illustrated by Barnes et al. (1970). While European specimens of *Oistodus lanceolatus* Pander may be hyaline (Lindström, 1971), those of the Midcontinent Province have variable, and commonly large, proportions of white matter. Ethington (1972, pp. 21–22) noted this taxonomic problem with *O. lanceolatus* and several other Lower Ordovician taxa. Even the holotype of *Scolopodus cornutiformis* is hyaline while the paratypes possess some white matter (D. J. Kennedy, pers. comm., 1974). Considerable work remains to be accomplished in conodont ultrastructure before these anomalous relationships are resolved, particularly to determine which features are phenotypic and which are genotypic.

Summary

The ultrastructure is described for six form species representing two form genera of Lower and Middle Ordovician conodonts belonging to the Acanthodontinae. Several features common to all specimens are noted. The elements are comprised of numerous cone-in-cone lamellae which originate from an initial growth point at the tip of the basal cavity. The lamellae extend apically to form the cusp and basally to outline the basal cavity. It is demonstrated that after the maximum width of the cusp is attained, material is secreted only in those directions necessary to elongate the cusp, or to deepen the basal cavity. Maximum growth in a preferred direction is accomplished in one or more of the following ways: (1) by thickening of the lamellae through an increase in the number of the crystallites, as illustrated by *Ulrichodina prima* and *Scolopodus cornutiformis*; (2) by thickening the lamellae through reorientation of the crystallites within the lamellae such that the c-axis of the crystallites is perpendicular to the interlamellar species, as illustrated in *S. gracilis*; (3) by increasing the width of the interlamellar spaces, as illustrated by *S. gracilis*; and (4) by the addition of partial lamellae to isolated areas of the cusp, as shown by the termination of lamellae in *S. emarginatus, S. gracilis*, and *S. quadraplicatus.*

Each lamella is formed of innumerable hexagonal crystallites. The c-axis of the crystallites is oriented approximately parallel to the direction of maximum growth in the element. The outer surface of the crystallites, i.e., the 1010 face, outcrops on the surface of the element to form a finely striated ornamentation.

Few specimens have a smooth circular or ovoid cross-section, but rather display a prominent ridge-groove system. This further ornamentation is shown to have originated in one of two ways: either by the addition of partial lamellae preferentially at one or more locations on the cusp, as illustrated by *S. quadraplicatus*, or by thickening of the lamellae in specific areas, as seen in *S. cornutiformis* and *U. prima.*

All the studied specimens have developed white matter by the reorganization of hyaline (lamellar) material. This reorganization process proceeds through an intermediate stage producing incipient white matter.

Several other structures of note are restricted to individual species within the study. The most notable of these are: radial lamellae, keels, a longitudinal furrow, and the various specialized concentrations of white matter. All of these features have one thing in common: they are all lineations oriented in a longitudinal direction. This may suggest that they are all specialized adaptations designed to strengthen the conodont, the cusp being stressed to counteract forces applied from a vertical direction. Furthermore, it may be noted that all specimens are slightly to strongly recurved in the posterior-anterior plane, a feature which suggests a necessity to strengthen the element in that direction.

Acknowledgments

The continuing financial support for conodont studies from the National Research Council of Canada is gratefully acknowledged by C. R. Barnes.

Literature Cited

BARNES, C. R.

1974 Ordovician conodont biostratigraphy of the Canadian Arctic. *In*: Aitken, J.D. and D. J. Glass, eds. Canadian Arctic geology. Calgary, Geological Association of Canada—Canadian Society of Petroleum Geologists Special Volume, 221–240.

BARNES, C. R. and L. E. FÅHRAEUS

1975 Provinces, communities, and the proposed nektobenthic habit of Ordovician conodontophorids. Lethaia, 8: 133–149.

BARNES, C. R. and M. L. S. POPLAWSKI

1973 Lower and Middle Ordovician conodonts from the Mystic Formation, Quebec, Canada. J. Paleont., 47 (4): 760-790.

BARNES, C. R., C. B. REXROAD, and J. F. MILLER

1973 Lower Paleozoic conodont provincialism. *In* Rhodes, F.H.T., ed. Conodont paleozoology. Spec. Pap. Geol. Soc. Am., 141: 157–190.

BARNES, C. R., D. B. SASS, and E. A. MONROE

1970 Preliminary studies of the ultrastructure of selected Ordovician conodonts. Life Sci. Contr., R. Ont. Mus., 76: 1–24.

1973 Ultrastructure of some Ordovician conodonts. *In* Rhodes, F.H.T., ed. Conodont paleozoology. Spec. Pap. Geol. Soc. Am., 141: 1-30.

BARNES, C. R., D. B. SASS, and M. L. S. POPLAWSKI

1973 Conodont ultrastructure: The Family Panderodontidae. Life Sci. Contr., R. Ont. Mus., 90: 1–36.

BARNES, C.R. and M. F. TUKE

1970 Conodonts from the St. George Formation (Ordovician), northern Newfoundland. Bull. Geol. Surv. Can., 187: 79–97.

BERGSTRÖM, S. M.

1973 Ordovician conodonts. *In* Hallam, A., ed. Atlas of palaeobiogeography. Amsterdam, Elsevier Scientific Publishing Company, pp. 47–58.

ETHINGTON, R. L.

1972 Lower Ordovician (Arenigian) conodonts from the Pogonip Group, central Nevada. *In* Lindström, M. and W. Ziegler, eds. Symposium on conodont taxonomy. Geologica et Palaeontologica Sonderband 1: 17–25.

HASS, W. H. and M. L. LINDBERG.

1946 Orientation of the crystal units of conodonts. J. Paleont., 20(5): 501–504.

LINDSTRÖM, M.

1955 Conodonts from the lowermost Ordovician strata of south-central Sweden. Geol. För Stockh. Förh., 76(4): 517–603.

1964 Conodonts. Amsterdam, Elsevier Publishing Company, 196pp.

1970 A suprageneric taxonomy of the conodonts. Lethaia, 3(4): 427–445.

1971 Lower Ordovician conodonts of Europe. *In* Sweet, W. C. and S. M. Bergström, eds. Symposium on conodont biostratigraphy. Mem. Geol. Soc. Am., 127: 21–61.

LINDSTRÖM, M., R. A. MCTAVISH, and W. ZIEGLER

1972 Feinstrukturelle Untersuchungen an Conodonten 2. Einige Prioniodontidae aus dem Ordovicium Australiens. Geologica et Palaeontologica 6: 33–37.

LINDSTRÖM, M. and W. ZIEGLER

1971 Feinstrukturelle Untersuchungen an Conodonten 1. Die Überfamilie Panderodontacea. Geologica et Palaeontologica, 5: 9–17.

MÜLLER, K. J. and Y. NOGAMI

1971 Über den Feinbau der Conodonten. Mem. Fac. Sci. Kyoto Univ., Geol. Mineralog. Ser., 38(1): 1–87.

PIERCE, R. W. and R. L. LANGENHEIM, JR.

1969 Ultrastructure in *Palmatolepis sp.* and *Polygnathus sp.* Bull. Geol. Soc. Am., 80(7): 1397–1400.

1970 Surface patterns on selected Mississippian conodonts. Bull. Geol. Soc. Am., 81(11): 3225–3236.

PIETZNER, H., J. VAHL, H. WERNER and W. ZIEGLER

1968 Zur Chemischen Zusammensetzung und Mikromorphologie der Conodenten. Palaeontographica, 128(A) 4–6: 115–152.

Fig. 1 A-F. SEM micrographs of *Scolopodus cornutiformis* Branson and Mehl.

A. Lateral view. Note the distinct sharp-edged costae but smooth anterior margin, × 180. ROM 30576

B. Lower view. Note the lamellae outcropping in the basal cavity, and costae on the posterior margin, × 180. ROM 30577

C. Transverse section one-third up from the basal margin and perpendicular to the growth axis, showing the growth axis (hole) and the concentric nature of the lamellae, × 430. ROM. 30578

D. Enlargement of Fig. 1C. Note the fusion of the lamellae across the interlamellar spaces and partial recrystallization of hyaline material to white matter (right), × 1700.

E. Transverse section at mid-length, showing the concentric nature of the lamellae, and prominent coarse costae, × 380. ROM 30579

F. Enlargement of Fig. 1 E. Note the internal representation of the surface costae by the sharply flexed lamellae, × 760.

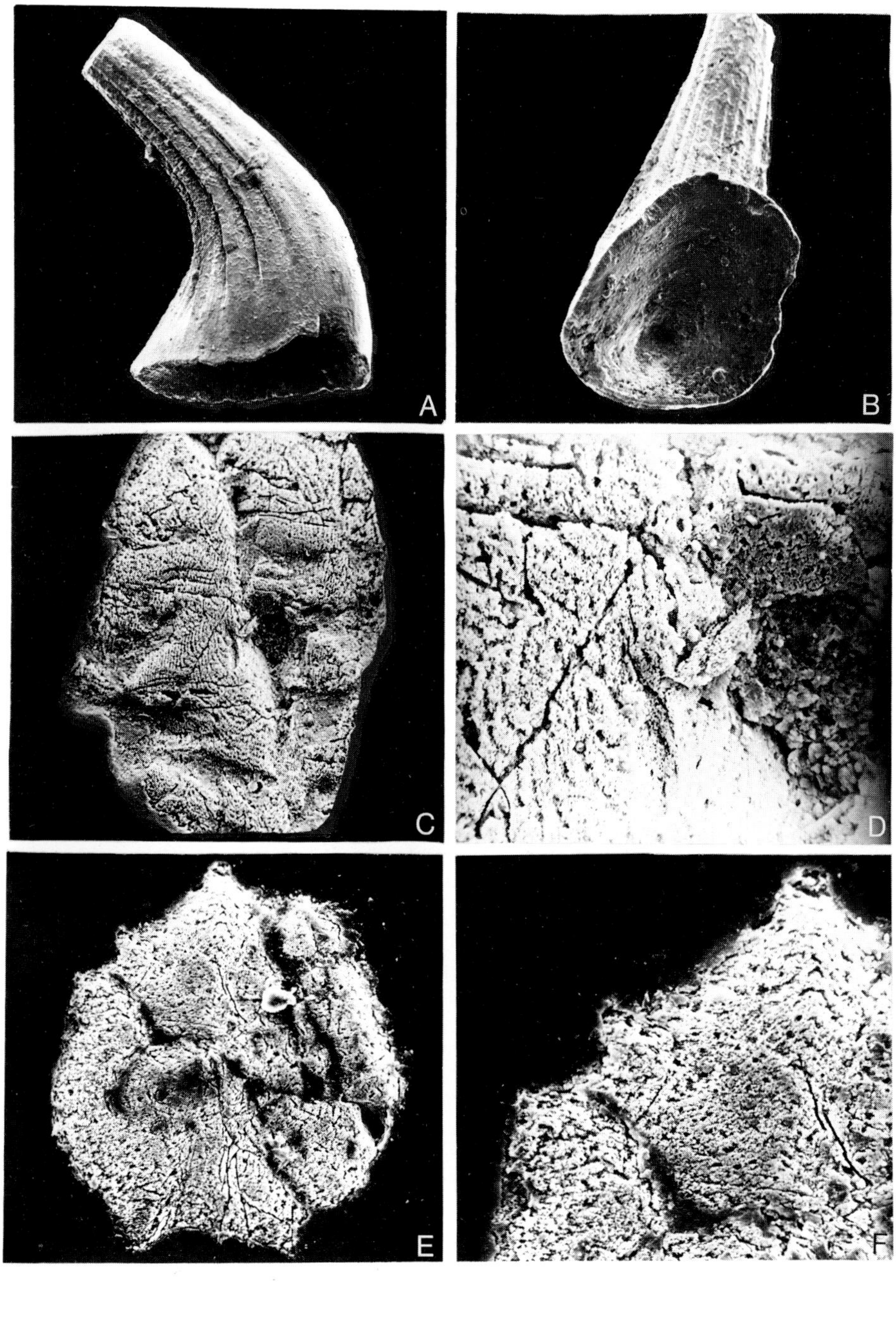
A
B
C
D
E
F

Fig. 2 A. *Scolopodus multicostatus* Barnes and Tuke, lateral view. Note the fine striations on the surface, × 88. ROM 30591

B-F. *Scolopodus emarginatus* Barnes and Tuke.

B. Lateral view. Note the fine surface striations and posterior groove (arrow), × 240. ROM 30585

C. Longitudinal section. Note the orientation of lamellae, the growth axis (arrow), and basal cavity, × 160. ROM 30583

D. Enlargment of the basal margin of Fig. 2C, showing the addition of lamellae perpendicular to the direction of growth and their restriction to the basal area, × 940.

E. Oblique cross section about $\frac{2}{3}$ up from the basal margin. Note the longitudinal furrow (arrow), × 750. ROM 30584

F. Enlargment of upper half of Fig. 2E. Note the internal continuation of surficial striae, × 1000.

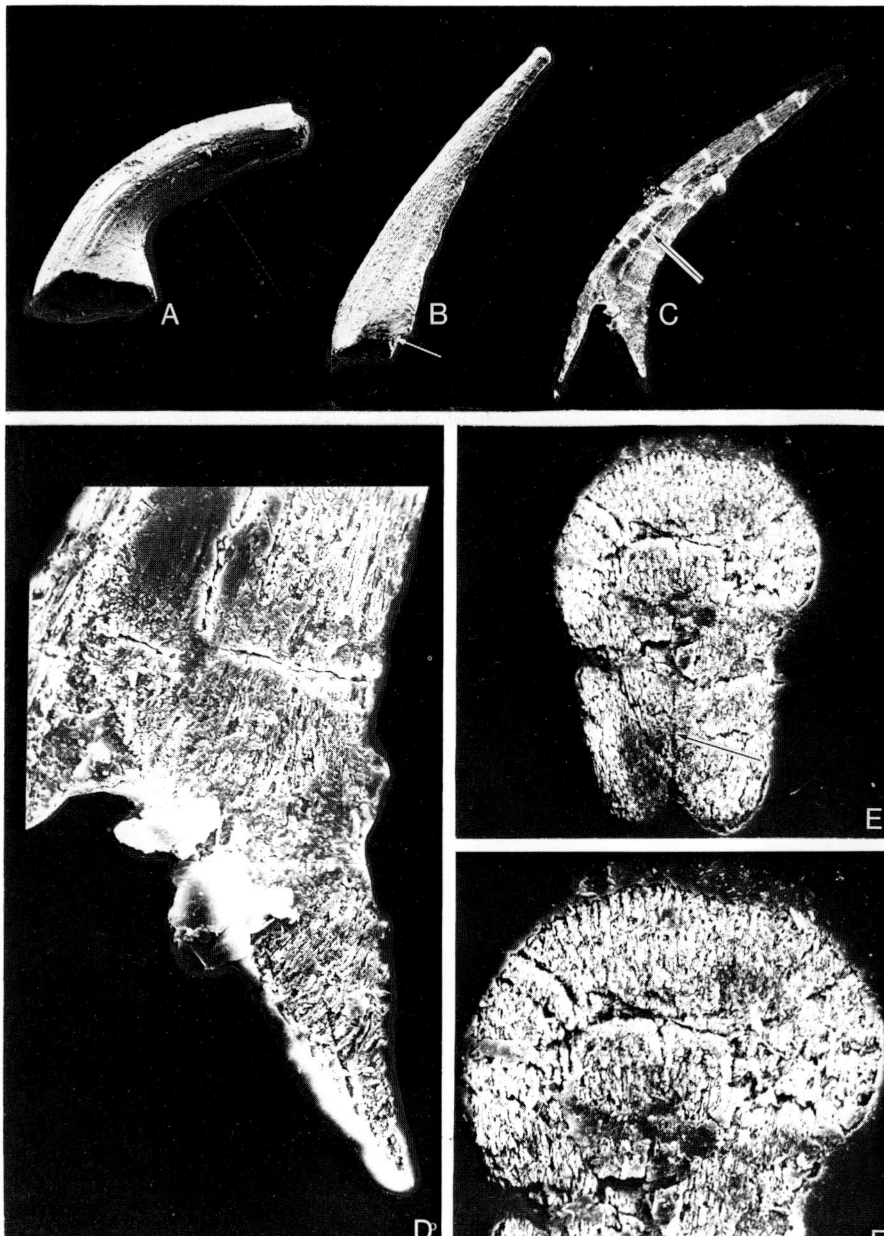
A
B
C
D
E
F

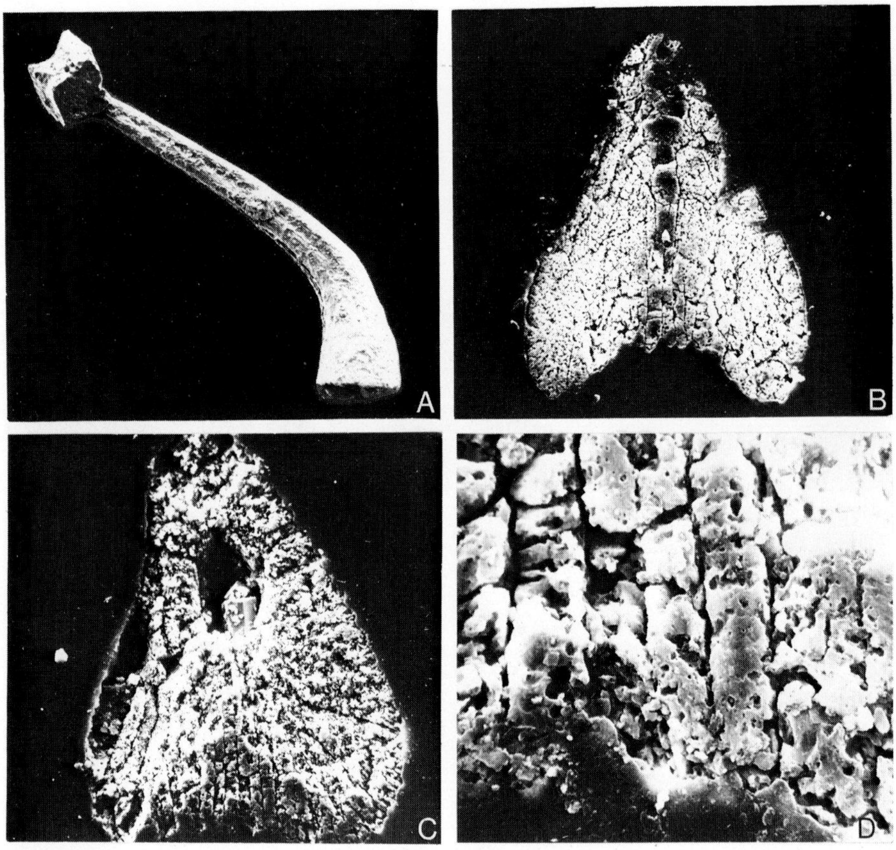

Fig. 3 A-D. *Ecolopodus gracilis* Ethington and Clark.

A. Lateral view. Note wide, broad carinae and fine, indistinct striations, × 185. ROM 30587

B. Transverse section through *S. gracilis*, showing a band of white matter extending from the posterior groove to the anterior edge, and the internal extension of coarse striae from the posterior groove (as radial lamellae), × 700. ROM 30588

C. Transverse section, lower in the same specimen illustrated in Fig. 3B, showing the enlarged growth axis (hole) and radial lamellae near the posterior margin, × 550.

D. Enlargement of Fig. 3C, showing radial lamellae composed largely of incipient white matter and surficial outline of the coarse striae, × 2600.

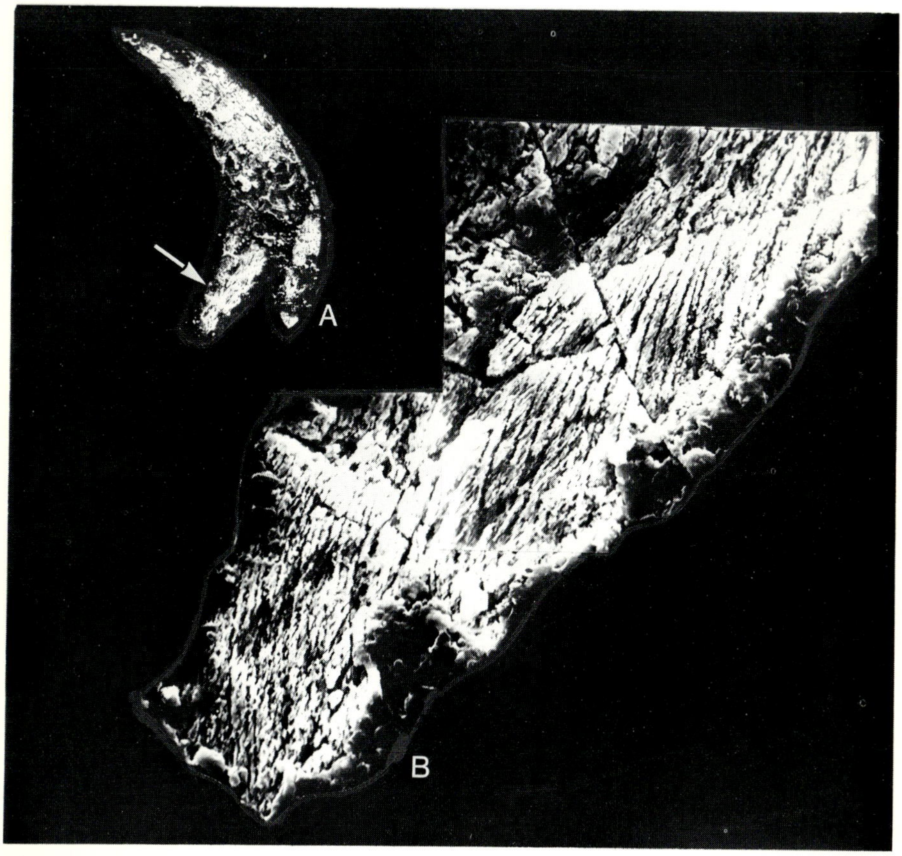

Fig. 4 A-B. *Scolopodus gracilis* Ethington and Clark. ROM 30589

A. Longitudinal section, × 225.

B. Enlargement of Fig. 4A (arrow) showing: the progressive extension of the posterior basal margin through lamellae added to that region only; the orientation of lamellae oblique to the direction of growth; and the oblique orientation of crystallites within the lamella, × 1750.

Fig. 5 A-E. *Scolopodus quadraplicatus* Branson and Mehl.

A. Lateral view. Note the characteristic prominent costae and superimposed fine striations, × 150. ROM 30595

B. Transverse section at mid-length, showing concentric growth pattern and the termination of lamellae into grooves with addition of lamellae only to ridge areas, × 440. ROM 30596

C. Longitudinal section, × 115. ROM 305797

D. Enlargement of upper part of Fig. 5C, showing the orientation of crystallites, the white matter in the tip, and the truncation of lamellae at the margins of the specimen indicating secretion of lamellae only in the distal area, × 460.

E. Enlargement of Fig. 5D (arrow), showing the transition from lamellar material to white matter in the central zone (left to right), × 5400.

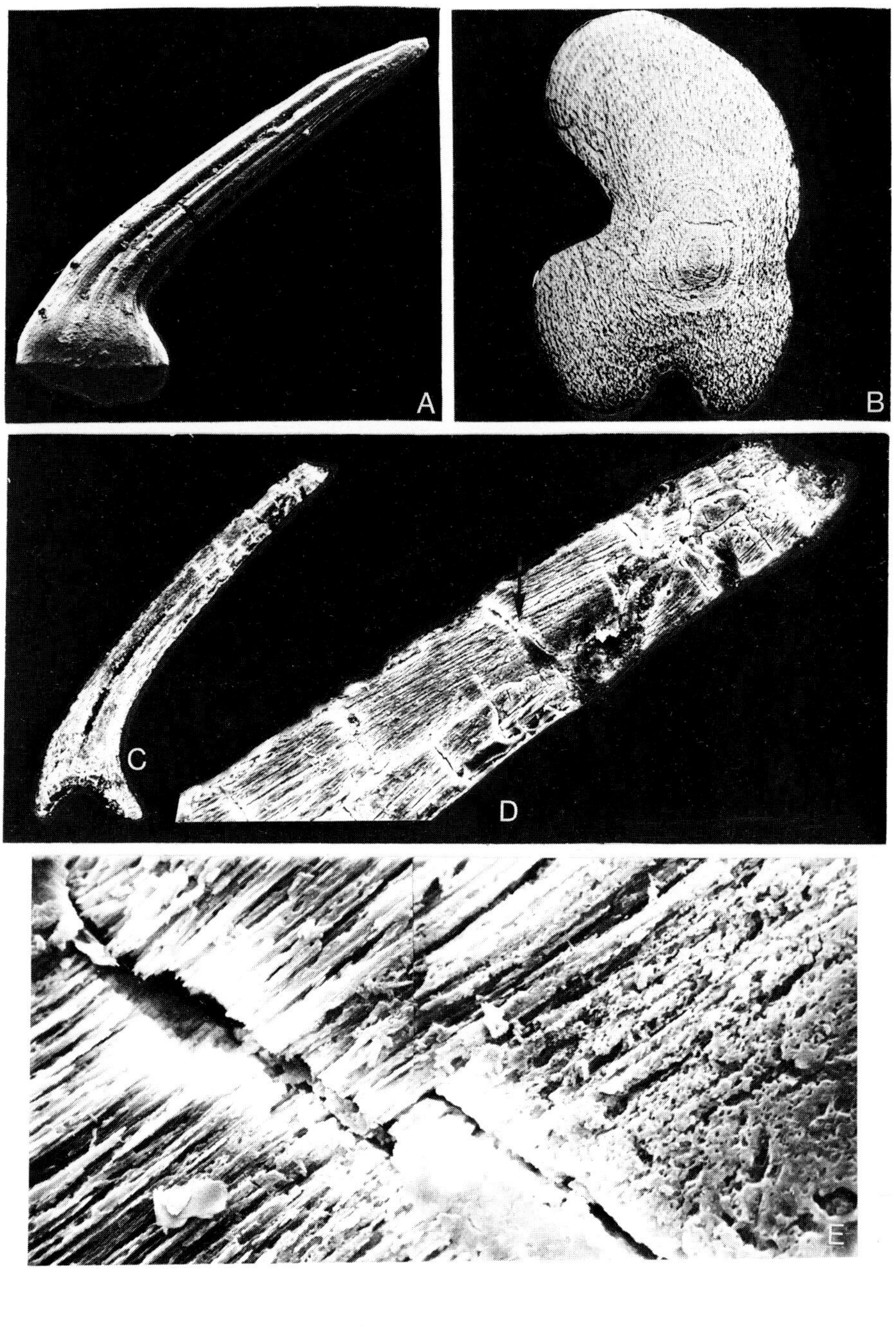
A
B
C
D
E

Fig. 6 A-E. *Ulrichodina prima* Furnish.

A. Lateral view. Note fine striae and posterior keel, × 150. ROM 30607

B. Enlargment of Fig. 6A, showing disappearance of striae on ridges and their presence in grooves, × 375.

C. Transverse section at mid-length, showing concentric growth lamellae and growth axis (arrow), ×235. ROM 30608

D. Enlargement of anterior margin of Fig. 6C, showing thin lamellae (average 1.2μ), × 940.

E. Enlargement of posterior of Fig. 6C, showing the thickening of lamellae to a maximum of 3.5μ at the keel. Note crenulations of lamellae, i.e., earlier ontogenetic surficial fine striae of early ontogenetic stages × 940.

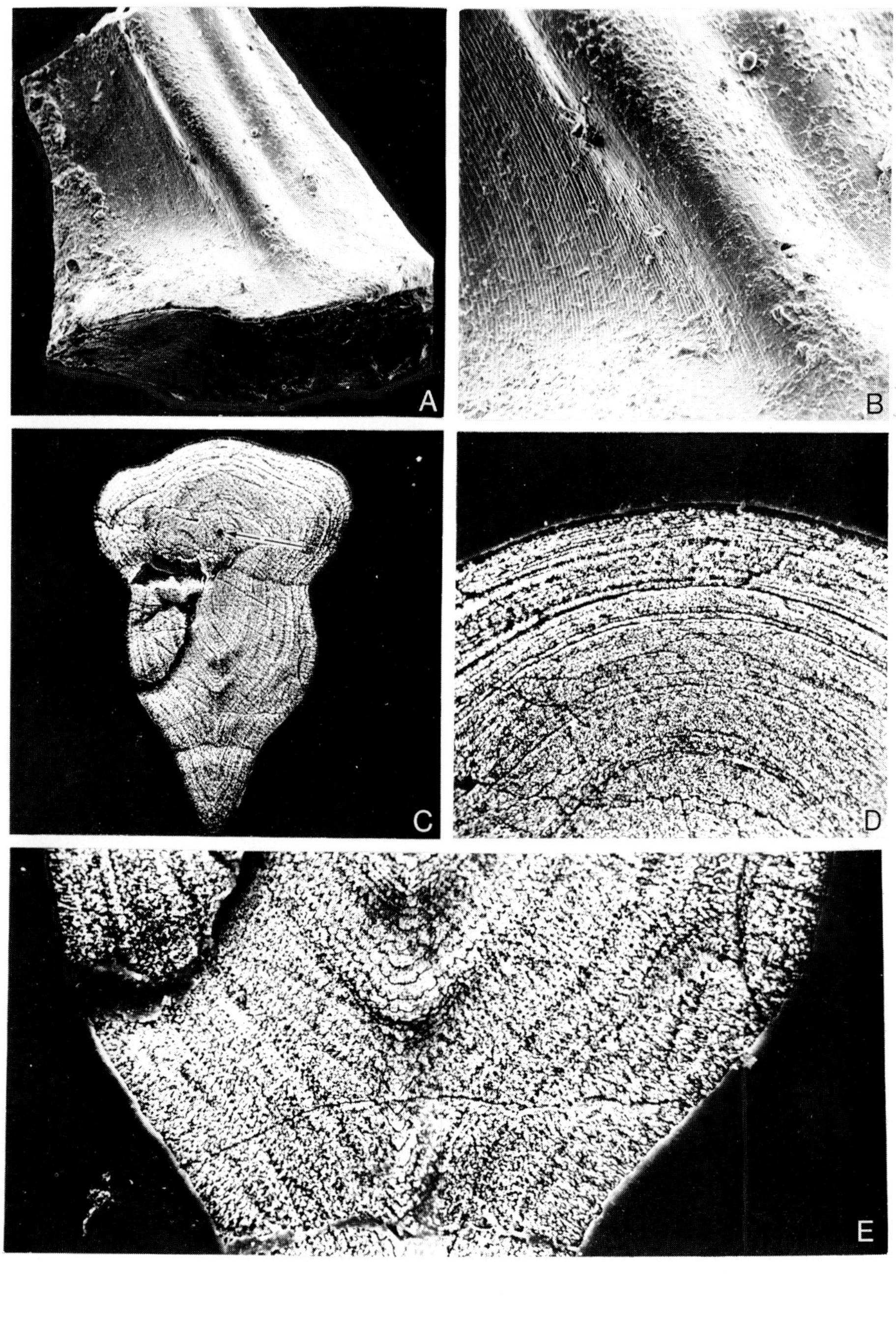
A
B
C
D
E